On a Cold Day

Dona Herweck Rice

Publishing Credits

Rachelle Cracchiolo, M.S.Ed., *Publisher*
Conni Medina, M.A.Ed., *Managing Editor*
Nika Fabienke, Ed.D., *Content Director*
Véronique Bos, *Creative Director*
Shaun N. Bernadou, *Art Director*
Valerie Morales, *Associate Editor*
John Leach, *Assistant Editor*
Courtney Roberson, *Senior Graphic Designer*

Image Credits: All images from iStock and/or Shutterstock.

Library of Congress Cataloging-in-Publication Data

Names: Rice, Dona, author.
Title: On a cold day / Dona Herweck Rice.
Description: Huntington Beach, CA : Teacher Created Materials, [2019] | Audience: K to grade 3. |
Identifiers: LCCN 2018029720 (print) | LCCN 2018029829 (ebook) | ISBN 9781493899104 | ISBN 9781493898367
Subjects: LCSH: Cold--Juvenile literature. | Temperature--Juvenile literature. | Weather--Juvenile literature. | Vocabulary.
Classification: LCC QC981.3 (ebook) | LCC QC981.3 .R4955 2019 (print) | DDC 551.5/25--dc23
LC record available at https://lccn.loc.gov/2018029720

Teacher Created Materials
5301 Oceanus Drive
Huntington Beach, CA 92649-1030
www.tcmpub.com

ISBN 978-1-4938-9836-7

Who can use some

 on this day?

cocoa

Who can use some

 on this day?

soup

Who can use some

 on this day?

coats

Who can use some

 on this day?

boots

Who can use some

 on this day?

scarves

Who can use some

 on this day?

hats

Who can use some

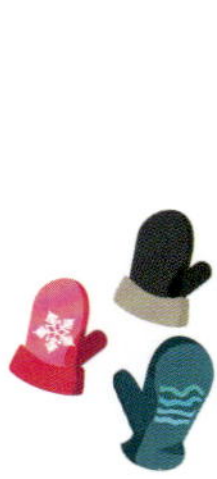

on this day?

mittens

Who can use some

 on this day?

blankets

Who can use some

 on this day?

socks

Who can use some

 on this day?

slippers

High-Frequency Words

New Words

can
day
some
use
who

Review Words

on
this